EDWARD COLE

The story of a
GEOLOGICAL
LIMESTONE
LANDSCAPE

THE EXTRAORDINARY LIMESTONE GARDEN AT
BOUNTY HALL, TRELAWNY, JAMAICA

EDWARD COLE

The story of a
GEOLOGICAL LIMESTONE LANDSCAPE

THE EXTRAORDINARY LIMESTONE GARDEN AT
BOUNTY HALL, TRELAWNY, JAMAICA

MEREO
Cirencester

Mereo Books

1A The Wool Market Dyer Street Cirencester Gloucestershire GL7 2PR
An imprint of Memoirs Publishing www.mereobooks.com

The story of a geological limestone landscape: 978-1-86151-788-3

First published in Great Britain in 2017
by Mereo Books, an imprint of Memoirs Publishing

Copyright ©2017

Edward L Cole has asserted his right under the Copyright Designs and Patents
A CIP catalogue record for this book is available from the British Library.

This book is sold subject to the condition that it shall not by way of trade or otherwise be lent, resold, hired out or otherwise circulated without the publisher's prior consent in any form of binding or cover, other than that in which it is published and without a similar condition, including this condition being imposed on the subsequent purchaser.

The address for Memoirs Publishing Group Limited can be found at
www.memoirspublishing.com

The Memoirs Publishing Group Ltd Reg. No. 7834348

The Memoirs Publishing Group supports both The Forest Stewardship Council® (FSC®) and the PEFC® leading international forest-certification organisations. Our books carrying both the FSC label and the PEFC® and are printed on FSC®-certified paper. FSC® is the only forest-certification scheme supported by the leading environmental organisations including Greenpeace. Our paper procurement policy can be found at
www.memoirspublishing.com/environment

Typeset in 12/18pt Century Schoolbook
by Wiltshire Associates Publisher Services Ltd. Printed and bound in Great Britain by
Printondemand-Worldwide, Peterborough PE2 6XD

PREFACE

The settlement at Bounty Hall, Trelawny, Jamaica is the site of a unique and extraordinary geological limestone landscape partially hidden by nature until it was discovered a few years ago. These rock formations were created by the solidification of soluble rocks created by volcanic activities. There are formations of solidified rock and vast arrays of loose rocks and fine particles of stones. They have now been used to create an artful range of displays as the basis of the Limestone Garden. This book has been written to inspire the reader and reach a wider geological audience.

This book is about the birth of a landscape and how it has evolved over a number of years. It is about all that it represents, what lies on it and around it, the wider surroundings and the elements that control it.

CONTENTS

Acknowledgments
Introduction

1	The Stone Garden Project	1
2	The geological background	6
3	The beautiful West Indies and the Parish of Trelawny	9
4	Developing the site	16
5	Exploring the Stone Garden	20

ACKNOWLEDGEMENTS

I wish to thank the following:

The **University of the West Indies**, Mona, Kingston, Jamaica.

All the people I have spoken to at the **Geographical Geological Department**, University of the West Indies, Mona, Jamaica regarding the white limestone landscape found on my property in Trelawny. I explained what its significance was and was impressed with their response. Although we were not able to meet I took specimens to the Natural History Museum, London.

The helpful staff at the **Natural History Museum**, London, for responding to my request to analyse the limestone specimens I submitted. I wish to say a very special thanks to the member of staff who wrote the report for her support in providing me with limitless information and exposing me to more information than I could ever have expected.

Kiera Batty (student), for her technical input towards this project.

Jon Batty (IT engineer), for his IT and communications help and his continued support of this project.

Wayne-Anthony Cole (Performing Arts), for providing me with communication materials and supports of this project.

Joan Clarke-Cole, for her support in the development of this project.

Ionie Jackson, for her continued support through the development of this project, and in particular for allowing us to photograph her for the cover.

Zaria and Zaire Beckford (students), the two first two youngsters to have their photographs taken on the site (pictures inside).

Ewatt and Delores Green, for their Barbados adventure tours, which have allowed me to take pictures of the magnificent countryside and of course Harrison's cave. The experience was most invaluable.

Kamron Shahzad (IT Points Services), for his outstanding IT services in supporting the development of this project.

Marcia Johnson BA (Hons Dp) LA, for her most welcome input and support. Her artistic suggestions to maximise the development of the pictures was most welcomed.

Donald and Barbara (grounds staff), for their continued loyalty and support in maintaining the grounds of the landscape.

Photograph of the author by Kristy Verdi of Verdi Studios (East London).

Edward Legister Cole

INTRODUCTION

I am a Jamaican with dual British nationality, and I was born in Albert Town, Trelawny, Jamaica in 1940. Since 1974 I have owned a property at Bounty Hall, near Wakefield, Trelawny. During the course of working on the property I discovered the hidden landscape which forms the subject of this book.

In 2011 I was working on the construction of a water management system when a heavy downpour of rain drew my attention to an exposed rock formation. On closer investigation, I was intrigued to discover that the area was the site of a limestone landscape of extraordinary variety. Everywhere I looked I found limestone formations in striking shapes.

This landscape was created by a volcanic explosion some tens of millions of years ago, and the blast deposited loose rocks over a wide surrounding area. Erosion of the soil is constantly exposing more of these rocks. The photographs in this book show how the rocks have emerged from the ground in loose piles.

I am still overjoyed by this discovery made in my own back yard. This landscape means a lot to me in many ways. It's the first piece of land I have been able to call home. In Jamaica we plant trees and watch them grow, transforming the skyline and producing fruit. The rocks and the trees are two of the elements that make this landscape so special and personal to me.

1
THE STONE GARDEN PROJECT

Once I was able to identify how the stones and rock formations had arisen, I began to think about how I could expose and display them for others to see. To gain a better understanding of what I had found, I engaged someone to remove the soil and let nature take its course and clean up the stones. As well as improving on the existing rock piles I discovered that I could make them into an added theme attraction in addition to the construction of the engineering water management system.

Once the discovery was made, everything became possible. The land was acquired in 1974 and the rock became the foundation on which my house was built. I set out to create a garden designed to bring inspiring and interesting features to the landscape by way of designed paved terraces with art form outlines and other features. From conception to fruition this project became part of me, from design sketches to implementation.

One of the interesting aspects of the site is that it lies on the geological fault line that runs through the Americas and under the Caribbean. While sitting in the house you feel slight earth tremors on a regular basis, and it is not unusual to hear the sound of the glass louvre blades rattling.

I soon realised that a water control system was needed to divert heavy storm water away from the lawn and terraces to minimize erosion. I put a great deal of thought into constructing a system that was practical and at the same time artistic. This posed some challenges, because I was unsure how it would affect the stone surfaces, so that phase was left to the last to be implemented.

The Stone Garden became one of the most challenging projects I have ever attempted. It has taken me into uncharted territory, with moments of highs and lows, and sometimes I have wanted to call the whole thing off. There were many setbacks. After two years of drawing up plans, I went to the bank to withdraw the first quantity of cash I needed for the project and was robbed on my way back. All the money was stolen.

Labour issues proved challenging at times as well, although my experience in the building industry helped here. But at moments like these I knew I would be disappointed with myself for letting myself down so badly. I have never walked away from a challenge where the end result could potentially be rewarding. There were so many times of uncertainty with lots of question marks, yet I persevered.

During the moments of despair, I tried to concentrate on the areas that mattered most. My day job as a craftsman has given me the right "building blocks" to start from on. Despite all the adversity, the high blood pressure and the worry, it's a journey that has given me immense pleasure and has proved educational in so many ways.

CHALLENGES

Designing the garden was simple enough, but executing the project has posed a few challenges. Once the work began some of those fears went away. We got through it eventually and were able to finish the job. It has transformed the neighbourhood and I have had many positive reaction from some visitors.

The project has proved to be more exciting than I expected. It helped that the site was able to offer everything I needed to make a success of it without buying in materials from elsewhere.

My ultimate ambition is that the project will inspire others to do likewise to improve their own spaces. I hope schools will use the site for field studies. Already one of my immediate neighbours has begun work to expose the rock face on his own property.

A PHOTOGRAPHIC RECORD

When I was working in the building trade, taking pictures was part of the surveying procedure as the pictures can be used as reference for before and after images. In fact I enjoy taking photographs for pleasure as well as to record features for work purposes. The photographs taken at Bounty Hall span over thirty years spent on the site, and I now have a library of several hundred images. I have spent thousands of hours poring over these pictures and checking the details of each of them to look for any hidden secrets that I might have overlooked before. Nearly all the pictures have been reworked many times over and work is still in progress as I write.

When I started this project I was entering the unknown, and my photographic equipment was not up to the job, so I had to replace it. Now I am on my third set of equipment and very much up to date. I have been learning on the job, but that's nothing new as my entire working life has been a learning process, which has equipped me to overcome some of the challenges I encountered. My current equipment has allowed me to break new grounds with the processing of the pictures.

The aim of this process is to present this project in

an exciting, constructive and informative way. Many people would have to travel long distances to view such a location.

2

THE GEOLOGICAL BACKGROUND

Through consulting experts at the Natural History Museum in London after my discovery, I established that the rocks were from the White Limestone Group, which cover about two thirds of the land area of Jamaica and largely appear as the limestone formation known to geologists as karst. They are Oligo-Miocene in date, having been formed between approximately 50 and 10 million years ago. There are few if any other places in the world where you can see such a rich array of examples of nature's work in carving various kinds of rock formation from limestone.

The West Indies are well documented as having evolved after a massive volcanic eruption, creating high mountain ranges here in Jamaica and in Cuba. The faults that produce these eruptions are still active, and Jamaica has had many more volcanic eruptions over the centuries. The land was used by British settlers from 1655 and most of their activities were for agricultural purposes, mainly for sugar cane plantations. Those operations are documented and I have one such document to hand which explained that they acquired the services of Africans, Chinese and Indians to be the chief workers for the sugar cane plantations and to build factories to facilitate sugar and rum manufacturing.

The volcanic push-up that created the formations is about a square mile in extent, with high and low areas. Part of the formations lie across my boundary on a neighbouring property, but my neighbours had no idea what they were until I explained. Only a very limited area is exposed. The rest is covered with trees and bushes. The area is extensively farmed and there are dwellings and farm buildings all over the area. The exposed surfaces on both sides extend to about 2000 square metres, or approximately a quarter the size of a football pitch.

During those early years the land would have been covered with woodland, and on this site in particular the surface was littered with loose stones and rocks alongside the geological formations. Some stone was used for dry walling, while the smaller ones were collected in piles which are evident today – some of them can be seen in the pictures.

History shows that the land was first occupied by three groups of people. First were the Caribs and the Arawak Indians, then came Christopher Columbus, of

Italian origin, supported by the Spanish Government. He arrived in 1492, followed by the English, who took control in 1655.

It's interesting that this site hold so much history, now that the sectors are assessed and the big picture emerges. All over Jamaica there are many historical sites, some on a much bigger scale, but few are more interesting.

THE KARST LANDSCAPE

Karst is a landscape formed from the dissolution of soluble rocks, including limestone, dolomite and gypsum. Rainwater becomes acidic when it comes into contact with carbon dioxide in the atmosphere and the soil. As it drains into the fractures in the rocks, the water begins to dissolve away the softer rock, creating a network of passages. Over time water flowing through the network continues to erode and enlarge the passages, which allows the system to transport increasingly large amounts of water. This process of dissolution leads to the development of caves, sinkholes, and sinking streams.

The word 'karst' was borrowed from a German word in the late 19th century which was named after Carso, a limestone plateau surrounding the City of Trieste in the northern Adriatic (nowadays the border between Slovenia and Italy).

3

THE BEAUTIFUL WEST INDIES AND THE PARISH OF TRELAWNY

Trelawny is the fifth largest of the fourteen parishes of Jamaica, lying between St Ann and St James on the north coast of the island. Much of Trelawny is rugged and mountainous. Most of the parish is several hundred feet above sea level, and it provides most of the country's rainfall as well as its plants and minerals. It is a limestone region with many features typical of limestone such as caves, sinkholes and underground rivers.

The landscape goes back to the days of the early settlers, the Spaniards, who arrived after Christopher Columbus. The English arrived in 1665, fighting the Spaniards and driving them out. They used the land for agricultural purposes, mainly for sugar cane plantations. My land at Bounty Hall was once part of the property of the Hampden sugar estate, which still exists today, though now they only manufacture rum and grow other farm products.

This land became available through acquisition under the Idle Land Redistribution Programme of the then Prime Minister Michael Manly. There is more major geological interest in the bordering five parishes. The parish is bordered by rainforest in all four neighbouring parishes.

This is the Duns River falls at Ocho Rios, Jamaica, one of our island's most famous tourist attractions. It is close to the town centre and the main road crosses over just before it reaches the sea. The falls attract a wide range of visitors. Climbing them is the experience of a lifetime.

This scene was photographed from the grounds of the Anglican Theological College on the Atlantic coast of Barbados, when the ocean was at its calmest.

Harrison Cave, Barbados, another of the wonders of the world the West Indies has to offer. Looking at it gives the impression that time stands still. This formation is very much active and growing.

A TRIBUTE TO ZARIA AND ZAIRE

Zaria and Zaire Beckford

I wish to congratulate these two Jamaican youngsters, who were the first children to view the newly-discovered rock formations. My ambition for these historical rocks is for them to be used for school field studies. In 2012 I took specimens to the British Natural History Museum London for analysing, and the findings I received were very interesting. Zaria had an interest in geography, so I gave her the specimens and invited her to take them to her school to be used in teaching geography.

Congratulations to Zaria for her great interest in my discovery. I hope there will be many more to follow.

I have named this plant Rubycilder. It is dedicated to my mother, who planted it in the 1990s. This picture was taken during an early tropical misty morning.

Views at sunrise. The tranquil scene above will only be there for a few minutes before the warmth of the rising sun disperses it. These are typical of the scenes the landscape has to offer. The camera has turned them into timeless moments. My son and I replanted it in the summer of 1980.

THE STORY OF A GEOLOGICAL LIMESTONE LANDSCAPE

As the sun dips below the horizon at the closing of the day, the treetops say farewell until tomorrow.

The landscape of Trelawny goes back to the days of the early Spanish settlers after the arrival of Christopher Columbus. The English arrived in 1665 and drove the Spanish out. They used the land for agricultural purposes, mainly for sugar cane plantations.

This land was once part of the property of the Hampden sugar estate, which still exists today, although they now only manufacture rum and grow other crops. The land became available through acquisition by the then Prime Minister Michael Manly, under the idle land redistribution programme. When the land was prepared for agricultural use it was cleared of loose rocks and piled in several heaps, some of which remain in the same places today. The heavy weathered formations to the front of the picture were placed there in 1980 by my son and me.

4

DEVELOPING THE SITE

The name 'Stone Garden' reflects the fact that the many distinctive stones are such a strong feature of the garden. The first time a section of metamorphosis formation became noticeable was in 2011, during a heavy downpour of rain during a water management construction project, and this led to further investigation of the site.

When I discovered the geological riches of the site, I was at first unsure what to do to improve it. Once the discovery was made everything became possible. From conception to fruition this project quickly became part of me, through from design sketches to implementation. The landscape itself with its sloping surfaces contribute to the designs, as can be seen on these pages.

The creative design of the scheme reflects my experience in the woodworking industry, from cabinet making to working with Saville Engineering Construction. Some of these elements of creativity influence the arrangements of the way the pictures are brought together. In some instances, they may appear repetitive, but they are constructed differently to demonstrate that the same materials have a versatility advantage of assembling as an example seen on page 19.

The theme is geological, and is complemented by the natural surroundings, from solid rock and pebbles and from shrubs to fully-grown trees. I have observed over the decades how they have shaped the skyline and now the grounds.

THE EMERGENCE OF THE SCULPTURE AND THE STORIES BEHIND IT

When Ionie Jackson and her children paid me an unexpected visit, I took the opportunity to take this picture.

Once the groundwork was completed I knew I had to do something else to attract interest from a distance in the form of a sculpture, but I had no idea how to accomplish this. Shortly thereafter I accidentally stumbled on a stone which was partly exposed. I had to remove it because it was in the way of another job, and there and then it provided the perfect answer to accomplish the sculpture. The lady in red in the picture, Ionie Jackson, paid me an unexpected visit along with the children, and it was the right moment in the right place to have this picture taken with Ionie in such a gracious posture.

Part of the house and grounds

The grounds of the garden are most interesting, comprising a diverse variety of rock formations - the photographs capture some of those elements.

At first I had no idea what could be responsible for the unusual variety of rocks, but I got the answers from research which showed that there must at some time in the past have been volcanic activity in the region, as shown on a geological map. It also shows the different types of limestone that exist in the region.

This picture shows a plant in the early stages of growth. The continuation of nature at work is making its presence felt and the rest is up to you. It is graciously appreciated because it makes a pretty picture.

5

EXPLORING THE STONE GARDEN

The Stone Garden looks different at each stage of the day as well as in different weather conditions. I never cease to find new views and new ways of looking at the land and the stones.

The entrance to the property

These pictures illustrate the different areas of the landscape, the immediate vicinity, the wider surroundings and how it affects the countryside. In the tropics, away from the towns, weather conditions are different from the cities. When you watch the Sun rise slowly over the horizon, the experience varies depending on the atmospheric conditions. I find the reflection of the Sun interesting when it filters through broken clouds as it slowly appears over the horizon.

At mid-morning the contrast in the garden is as its strongest. Sunrise brightens up the open spaces and filters through the branches of the trees. The outlines of the trees and branches create a transparency like no other. The backdrop highlights the white limestone terrace in real time, a contrast to that of a landscape painting.

During the night the warm air lifts the moisture from the ground, and as the sun rises the moisture turns to cloudy vapour and disappears with the rising temperature. As the mist rises it reflects the clouds and gives a silhouette effect through the branches and needles of the conifer trees which lasts for only a brief moment. To see this you have to be up at the break of dawn to see the transformation of the elements, because suddenly it's all over. You have to be a nature lover to make the observation in a meaningful way.

The morning dew settles on the vegetation like fine raindrops, and where there are plants with broad leaves like bananas the dewdrops settle in bubbles which glow by the reflection of the clouds, and it is quite an experience to see this.

When I was a youngster at school one of the songs we sang at the end of the day was 'Now The Day Is Over', because it was a church school which celebrated the teachings of the Bible. Those thoughts and sentiments still resonate with me today. I was born in the countryside, so we were surrounded by what nature has to offer. This landscape is rich in such offerings.

A bright sunny day outlines the shapes of the branches and the dense foliage of the trees, which appear just like a piece of artwork on a canvas. There are other times when there is moisture on the grass and shrubs at an angle which catches the sunlight and reflects it like a mirror. When there is a golden sunset the earth is like a receptacle waiting to receive the sun.

The colours reflect as if it has just been through a gold-plating process, and the whole neighbourhood turns to gold. You have to be in the right place at the right time with the camera to capture golden moments like these.

During the dry season you can see rain falling in the distance over the rainforest, while the ground where you are standing is parched and animals are dying. It is quite an experience to see this unfold in front of your eyes. If you then go into the area where it is raining, the vegetation is lush, while outside it the ground is bare.

The stones have now taken centre stage in the landscape, their presence dominating the scene. I was surrounded by stones in my childhood, but the only stone I saw when I moved to the UK was through my work in the building trade. Going back to Jamaica and having its native stones around me brings my life full circle. This time they have consumed me for all the right reasons. You cannot enter the property from any direction without the stones catching your imagination.

This is a part section of the front terrace, the sunlight shining on the stones

A truly remarkable collection of loose metamorphic rocks found among the large variety of shapes and sizes that makes up the collection. The formations remind me of watching a jeweller create a wax mould for jewellery making.

This picture shows the front terrace. Designing the Stone Garden began with this collection of stones, laid out to brighten up the rugged landscape.

This picture shows part of the water management wall which diverts storm water away from the house and terraces.

There are various fruit trees found in the garden, including this Ackee tree, the biggest one, which bears the fruit of Jamaica's national dish. There is also a mango tree, a banana plant and a coconut tree all jostling for space. These elements contribute to the diversity of the landscape.

Segments of plants and shrubs are carefully selected and crafted together to create an artful display, enhanced by the stones.

Below are some examples of the astonishing variety of stone formations found on the site. At every turn one encounters something different.

A truly remarkably piece of sculptured rock formation found in the limestone landscape.

THE STORY OF A GEOLOGICAL LIMESTONE LANDSCAPE

Above is one of the many fragmented fractures. The central area could not be more fitting. The formation of the fragments of the inner circle is one of the many surprises in nature's portfolio, complete with the shaping of the grass.

A classic piece of limestone landscape pavement crafted by nature from fractured rocks with the help of water over millions of years, and now exposed for us to see.

These are the largest loose rocks found on the site and the first to be used decoratively in the grounds. Those above were found over a wide area; I collected and brought them together to act as a buffer to the torrent of water during storms. They look as if they have lain there for centuries, but in fact they have only been there since 2011.

This stone emerged from the shadows to stand above the rest.

THE STORY OF A GEOLOGICAL LIMESTONE LANDSCAPE

The fragments of limestone vary considerably through the area, which adds to the significance of it all. The camera captures all the details of the different parts of this stone's composition.

A stone with an unusual formation seen at ground level.

Above is a most magnificent and unique rock, which depicts the form of a leaf on a tree. It almost gives the impression of creativity to a stone sculpture. The perimeter was designed to enhance the appearance of the stone while diverting heavy storm water away from the rock surfaces.

The structure of these rock formations had me puzzled for years and heavy rainfall led me to further investigation.

THE STORY OF A GEOLOGICAL LIMESTONE LANDSCAPE

These pictures show the variety of shapes in a small area of the landscape.

This shows a section of the grounds shortly after heavy rainfall which has changed the colour of the surfaces, while the sunset adds contrasting colours to the scene.

The golden sunset gives the stones a glowing effect which highlights even the smallest particles on the surface. It is quite remarkable to have this exposed to the elements.

This section of rock is one of the most interesting of all, and there is a story behind it. The house was built at the foot of the slope where this formation lies. I did not like it much because it gave little scope for planting. Because of the flow of heavy storm water against the house, I decided to build a storm water management system. The day after the wall was built there was a heavy downpour, and I stood on the veranda to see if the wall was fit for purpose. When the rain had stopped, I took a closer look and realised just how interesting the rock was.

That was the beginning of my discovery. I began to take a more detailed look at the rest and set about removing the soil over a wide area, revealing the landscape. These sections comprise various forms of solidified molten rock.

THE STORY OF A GEOLOGICAL LIMESTONE LANDSCAPE

This formation is different from the rest, and it caught my imagination the first time I saw it. All of these elements become the building blocks which have created the landscape, changing what was once just a rugged piece of ground into something that makes a striking difference to its surroundings.

The stones respond to the brilliance of the sunshine.

This rock, fractured in a chequered pattern, became the symbol of the landscape. It is quite remarkable how it is structured with near symmetrical formations. Soon after cleaning off the surfaces on a bright sunny day I went back, and I was standing gazing at it in admiration when I was struck by lightning! The shock of it knocked me to the ground, though fortunately I was unhurt. The lightning caused damage to properties in the surrounding area, so I had a lucky escape.

THE STORY OF A GEOLOGICAL LIMESTONE LANDSCAPE

This photo shows the structure of the solidified composition of this unusual rock enhanced in high resolution. The picture below shows the rock in normal lighting. It was found while clearing the site for the extended garden, partly exposed and lying in the middle of a footpath. While removing, it I discovered that it has unusual feathered markings, so I asked an archaeologist to take a look at it because I thought it might have been used by the early settlers of mankind, the Caribs. I named it the Pathfinder, and it is now a treasured stone.

These are different stones brought together to be displayed as part of the diverse collection of stones found in the grounds. There was a time when these stones meant nothing to anyone, and in fact were regarded as a nuisance, but time changes everything.

This is the scene when the sun filters though the branches of the trees as the shadows fall on the ground and brighten up the tops of the trees before it fades away over the horizon. These loose rocks are sculpted in the likeness of a dragon in the Chinese culture. Their origin goes back to the days of the arrival of the English in 1695 onwards to create sugar plantations. Today they are celebrated as a landscape of loose rock formations.

This is part of a rock formation with distinctive hidden features. The centre circle could have been caused by an escape of trapped air. The hole has been widened over the millions of years by the force of water passing over the surface.

THE STORY OF A GEOLOGICAL LIMESTONE LANDSCAPE

The loose rocks in these three photos are part of a wider collection of specimens found in the grounds of the garden. The surface is littered with stones which are revealed all the time by erosion of the soil. When the land was used for farming, the stones were a problem as they must have made for intensely hard work. Now some of these unique stones are celebrated as the creation of Mother Nature. Today they can be used artistically for anything you might possibly think they could be used for. The only limit is your imagination. I already have ideas for the next project.

This is another truly remarkable pattern of stone formation. In this case I was removing the soil around the borderline I was creating, and another mind-blowing surprise that appeared.

THE STORY OF A GEOLOGICAL LIMESTONE LANDSCAPE

Another unusual rock formation.

Here the grounds can be seen during early development. This picture was taken after the grass began to grow again, while the weather slowly changed the surface colour. It was after this process that the sculpture was erected and slowly the fractures began to get wider as the rain removed the soil. I have an interest in geology and try to have an understanding of its impact in all its forms on the planet. People travel all over the world to visit geological sites, and this one, being partly in step formation, provides easy access.

THE STORY OF A GEOLOGICAL LIMESTONE LANDSCAPE

Air trapped within this solidified rock formation has created voids as the rock cooled down, and as water has filtered through the openings over millions of years the voids have got larger. The image shows how complex and interesting this process is.

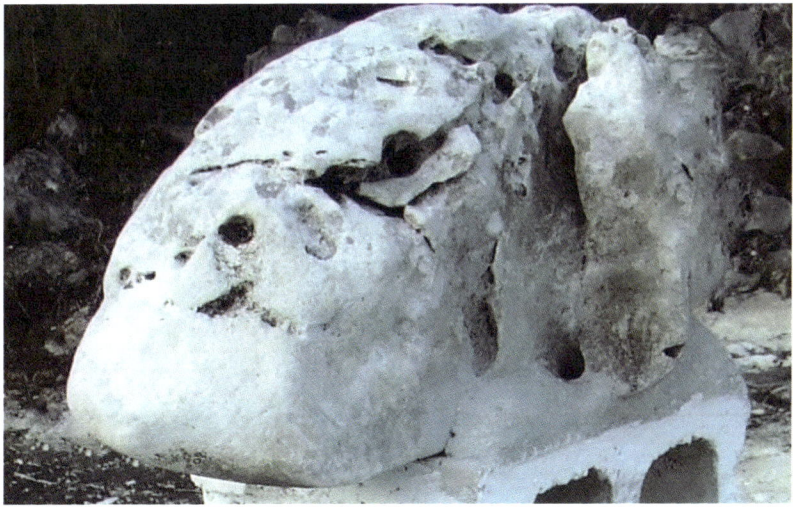

Above are some more of the treasured trophies found in the grounds of the landscape. The top picture shows the end of the water diversion system at the front. The stone mass below is known as the Focal Point and Head Stone.

Printed in Poland
by Amazon Fulfillment
Poland Sp. z o.o., Wrocław